Ludvigsen Library Series

PORSCHE SPYDERS
TYPE 550 1953-1956

Introduction by
Karl Ludvigsen

Iconografix

Iconografix
PO Box 446
Hudson, Wisconsin 54016 USA

© 2002 Ludvigsen Library Limited

All rights reserved. No part of this work may be reproduced or used in any form by any means... graphic, electronic, or mechanical, including photocopying, recording, taping, or any other information storage and retrieval system... without written permission of the publisher.

The information in this book is true and complete to the best of our knowledge. All recommendations are made without any guarantee on the part of the author or Publisher, who also disclaim any liability incurred in connection with the use of this data or specific details.

We acknowledge that certain words, such as model names and designations, mentioned herein are the property of the trademark holder. We use them for purposes of identification only. This is not an official publication.

Iconografix books are offered at a discount when sold in quantity for promotional use. Businesses or organizations seeking details should write to the Marketing Department, Iconografix, at the above address.

Library of Congress Card Number: 2002112573

ISBN 1-58388-092-5

03 04 05 06 07 08 09 5 4 3 2 1

Printed in China

Cover and book design by Shawn Glidden

Copyedited by Suzie Helberg

COVER PHOTO: Wilhelm Hild, in charge of the technical aspects of Porsche racing, was caught by Jesse Alexander stepping out of Porsche's 1,100-cc Type 550 Spyder at LeMans in 1955. The racer, pristine here, went on to win its class.

Book Proposals

Iconografix is a publishing company specializing in books for transportation enthusiasts. We publish in a number of different areas, including Automobiles, Auto Racing, Buses, Construction Equipment, Emergency Equipment, Farming Equipment, Railroads & Trucks. The Iconografix imprint is constantly growing and expanding into new subject areas.

Authors, editors, and knowledgeable enthusiasts in the field of transportation history are invited to contact the Editorial Department at Iconografix, Inc., PO Box 446, Hudson, WI 54016.

Porsche's Legendary Type 550 Spyder
by Karl Ludvigsen

In 1997 Britain's *Motor Sport* magazine asked me to name the greatest sports-racing car of all time. I chose the Type 550 Porsche for this honor. "Porsche's Type 550 Spyder achieved so much success in so many parts of the world for so many years," I wrote in its defense, "that it must be the greatest sports-racer ever. The Porsches dominated the 1 1/2-liter class and in race after race they punched well above their weight to rival vastly bigger cars.

"The greatest doesn't have to be the biggest," I added, in a vain effort to justify what I knew to be an eccentric nomination. I realized that most of my peers would be choosing the big cars, the Jaguar D-Types and Ferrari Testa Rossas, which hogged the limelight by winning the major races outright. In fact in the end they did elect a Porsche—the 12-cylinder Type 917—rather than the little four-cylinder 1 1/2-liter 550 Spyder. But the original 550, more than 100 of which were built between 1953 and 1956, must still rank as one of the very greatest sports-racers.

Porsche was still in its swaddling clothes as a car maker when the 550 series was born. It produced its first handfuls of cars in Austria in 1948 and '49 and in the latter year began its move back to its original home in Zuffenhausen, a suburb of Germany's Stuttgart. The first German-built Porsche car was made in the spring of 1950. It took the road in good time to be blessed by founder Ferdinand Porsche, who died at the age of 75 in January 1951. Thereafter his son Ferry, 42 years old in 1951, took charge of the family-owned company. Ferry inherited the outstanding team of engineers that his father assembled when he set up an engineering consulting office in Stuttgart at the end of 1930.

For Ferry and his team, motor sports were high in both interest and priority. They had designed and helped develop the great Auto Union racing cars of 1934 through 1937, and a racing-car project—for Italy's Cisitalia—put them back on their feet after the war. They weren't slow to return to the tracks. For racing Porsche used its Austrian-built Type 356 coupes, which were lighter with aluminum bodies. A solo entry in the 24-hour Le Mans race in 1951 came home with the 1,100-cc class victory, which was followed up with another class win in the world-famous French race in 1952. Porsche was in fact a plucky pioneer in returning a chastened Germany to international road-racing after the war years.

With competition heating up, Ferry and his team realized that they'd need to build special cars for racing as early as 1953. They had a convenient example to follow, that of Frankfurt's Walter Glöckler. A year older than Ferry, Glöckler was a great racing enthusiast who, as a VW dealer and Porsche distributor, could afford to indulge his passions. He did so with the help of his brilliant workshop chief, Hermann Ramelow. For the 1950 season they built a special tube-framed sports-racer that had a 1.1-liter Porsche engine mid-mounted, forward of its rear axles and behind the driver. Weighing only 980 pounds, the unique car took Glöckler to Germany's 1,100-cc class championship in 1950.

Impressed by the Glöckler team's competence and success, Porsche began cooperating with them in 1951, supplying 1.5-liter development engines to power new racers. In return the cars were known as Glöckler-Porsches and the Porsche name appeared on the body. This was good publicity for a new and still little-known make. Successful both in German racing and abroad—some were bought by Porsche's US importer, Max Hoffman—the Glöcklers provided a template for Porsche's own sports-racer, the Type 550.

Broadly speaking, the new car's type number meant that it was the 550th design project that the Porsche team had worked on since the company's founding. Numbers 1 through 6 hadn't been used—Porsche didn't want his first client to know he was the first—and many in the 400s weren't allocated. Project 549 was the design of a truck transmission for the Fuller Manufacturing Company of Kalamazoo, Michigan, while 551 was a three-speed gearbox design for a DKW car. Parked between these two was project 550, kicked off on October 27, 1952, for a sports-racing car.

Overseeing the project was a Wiesbaden-born engineer, Wilhelm Hild, who joined Porsche in 1951. Before the war chain-smoker Hild had been with DKW, where he worked on race-winning two-stroke motorcycles. Following the pattern set by Glöckler and Ramelow, he directed the building of two new Porsches using steel-tube ladder-type frames with mid-placed engines. As in the first Glöcklers, Hild turned the complete engine and suspension assembly through 180 degrees, which placed the transverse tube housing the rear torsion-bar springs at the extreme rear of the chassis. The resulting rear-suspension geometry wasn't ideal—nor was the rear frame structure—but with limited wheel travel it worked well enough.

Although conceived as a coupe for the lowest possible drag, one of the new 550s first raced as a roadster in the Eifel Races at the Nürburgring on May 31, 1953, in the hands of Helmut "Helm" Glöckler, Walter's cousin. He was victorious in a rainy event, thus ranking the 550 among those rare cars that were winners in their first race. Helm was still at the helm of 550-01 for Le Mans in 1953, where he co-drove with Hans Herrmann to second place in the 1.5-liter class a whisker behind 550-02, as the following photos will show and relate.

While these first relatively crude 550s were racing, Porsche's engineers were already at work on much-improved successors with better chassis and brand-new racing engines. Starting in 1952, Porsche's prospects blossomed with its winning of a half-million-dollar contract to design a new car for South Bend, Indiana's Studebaker. This, the Type 542, helped finance a new factory in Zuffenhausen as well as new sports-racers that could both promote the Porsche name around the world and generate money-spinning sales to wealthy enthusiasts.

From mid-1951 a new member of the Porsche team was Fritz Huschke Sittig Enno Werner von Hanstein, usually known simply as Huschke. Mad keen on motoring of all kinds, von Hanstein had raced both motorcycles and cars before the war. He was German hillclimb champion in a BMW 328 and co-drove the winning BMW coupe in the 1940 Mille Miglia. Huschke raced after the war as well, and was only ousted from the cockpit of one of the 550s at Le Mans in 1953 because Ferry Porsche thought he was better employed in the pits, managing the team. Heading both press relations and racing for Porsche, von Hanstein was the bright spark behind the successful commercialization of the company's on-track activities.

Another key player in the creation of the final 550 was Ernst Fuhrmann. A postwar addition to the Porsche staff, Fuhrmann was the son of a Viennese jurist. He earned an engineering doctorate from the Vienna Technical Academy and after the war survived as a watchmaker and locksmith in Austria. In 1947, at the age of 28, he joined the skeleton Porsche staff. Thanks to his academic background Fuhrmann was made the right-hand man of the team's chief theoretician, Josef Mickl. In the early 1950s he was assigned to engine development, working on valve gear, timing and camshaft design.

Small of stature and sharp-featured, Ernst Fuhrmann was shy of ceremony but his personal dedication and drive were strong. He was delighted to be offered the job of designing a new engine for the Type 550. This would be his chance to show what he could do. Backed by von Hanstein, who welcomed anything that would give his race and rally Porsches more power, Fuhrmann created the four-cam Type 547 four, an engine that was to be not only one of the most unusual but also one of the most successful in the annals of motoring and motor racing. A pure creation of the postwar era of Porsche engineering, Fuhrmann's engine was destined to give a new direction to engine research at Zuffenhausen. It led ultimately to the development of all Porsche's later air-cooled overhead-camshaft engines.

The new 550 to take the Type 547 was much improved. The transverse tube that held the rear torsion bars was now forward of the engine. This brought a more favorable rear-wheel steer effect and a reaction to braking torque that helped hold down the rear end of the car. The flexible steel arms which extended from the ends of the bars to the rear-wheel hubs had to be made 40 percent longer to allow enough room for the engine between the torsion-bar housing and the rear-wheel centerline. The longer arms had the effect of reducing the spring rate of the rear torsion bars, which were stiffened accordingly. Clothing this new chassis in 550-03 was a handsome new body from the pen of Porsche's body engineer/designer, Erwin Komenda. It was one of the prettiest and at the same time most purposeful shapes he was ever to originate.

With the creation of 550-03 in 1953, all the elements were in place that would lead to the production-model 550/1500RS or Spyder, as it was named, that was introduced at the beginning of 1955. This book's photos trace the development of both the engine and the car, showing that one of the early cars scored an amazing success in the 1954 Mille Miglia and that the new 550s built to compete at Le Mans in 1954 helped refine the model's design. We have as well a close look at the production-model 550 Spyder, one of the few racing cars to be delivered complete with chrome hubcaps.

A major contributor to the following pages is photojournalist Rodolfo "Rudy" Mailander. Born in Milan in 1928 of a German mother and Italian father, Rudy began writing and photographing cars in the late 1940s. From 1950 into 1955 he was a roving correspondent for the Swiss *Automobil Revue* and other periodicals, covering races, new-car launches and auto shows. A Porsche driver and enthusiast, Mailander was always welcome at the company's Zuffenhausen offices. Thanks to Rudy, whose negatives we hold at the Ludvigsen Library, we have almost all of this excellent record of the gestation and attributes of one of history's most successful and at the same time most attractive sports-racers. Other photos are from our own files.

Of the many 550 Spyders we were not able to show, one has iconic status. I refer to chassis number 550-055. This was a late-model Spyder, built in the summer of 1955, with the front brake-cooling inlets that had been added during its evolution. Brought to California, it was sold by Competition Motors for $6,900 to actor James Dean. Dean had raced a Porsche Speedster successfully and was eager to move up to a more potent car. He took delivery on September 21, 1955, and entered for the races at Salinas, California ten days later. While on his way there, driving on the road to get used to the car, Dean and his silver Spyder were struck by another car and the actor suffered multiple fractures and severe trauma. He was declared dead on arrival at the hospital in Paso Robles. His Porsche was a mangled wreck. Its engine was salvageable, but little else.

James Dean, whose debut role in *East of Eden* had brought great acclaim and whose last two films *Rebel Without a Cause* and *Giant* were released after his death, was the innocent victim of a banal road accident. Other Spyder owners were more fortunate, for this was one of the toughest and safest cars of its era. "Porsche's greatest attribute above all else," said Stirling Moss, who scored two 550 victories, "was their incredible, legendary reliability." Maintained according to the factory's meticulous instructions the Type 550 would rarely let its driver down.

An example was the 1954 1.1-liter 550 that scored three class wins in three starts in Europe. It was exported to the United States in time to compete in Sebring's 12-hour race in March 1955. There, driven by Fritz Koster and Paul O'Shea, it encountered opposition from a Lotus, which eventually retired, giving the small-bore Porsche another class victory. Then it was fitted with bigger pistons and cylinders to race as a 1.5-liter car for a season on the East Coast through 1956.

Acquired by the California team of John von Neumann, who cornered the market in Porsche's 1954 Le Mans cars, it was raced into 1957 by British expatriate Ken Miles. "We raised the compression," said Miles, "put the valve-spring pressure up, and substantially over-revved the little engine in three races. It hadn't been apart for three years and we knew it was due to blow any day, but it was running so nice and loose that we didn't want to tamper with it. About all we ever did was change plugs." Finally the engine shed a rod at Sacramento. "It had a perfect right to let go when it did," said Miles. That remark bespoke the affection that these cars engendered in their drivers at the time, and still do in the hearts of today's enthusiasts.

The first Porsche car created exclusively for competition was the Type 550 of 1953, a mid-engined chassis that could be raced either with or without its tapering roof. Its body was built by a Frankfurt firm, Weidenhausen, which had made bodies for special sports cars constructed and raced by Walter Glöckler.

The second of the two Type 550 coupes built in 1953, 550-02, at first had teardrop-shaped headlamps borrowed from a Ford Taunus. Before the car was raced these were replaced by conventional circular lamps. Its downsloping front deck promised good aerodynamic penetration.

The two Type 550 coupes were raced at Le Mans in 1953 with their hardtops in place, in spite of the excruciatingly high noise level they generated for their drivers. The enhanced streamlining helped them achieve 124 mph on the Mulsanne Straight with engines developing only 78 horsepower.

Porsche chassis 550-02 showed off its revised headlamps plus driving lamps while waiting for the start at Le Mans in 1953. In an astonishing success for a completely new model, this car won its Sports 1,500-cc class and finished 15th overall. Its sister was just behind it on the same lap.

A look under the rear deck of a 1953 Porsche 550 showed its tubular frame extending beneath its swing axles to a transverse tube at the extreme rear, containing the rear torsion bars. The same layout was used in the very first Porsche roadster built in 1948. Clutch operation was hydraulic and shock absorbers were tubular.

Twin-throat Solex carburetors fed fuel and air to the four horizontal cylinders of the hopped-up 1500 Super engine used in Porsche's 1953 Le Mans entries. The basic engine had pushrod-operated overhead valves and the air cooling that Porsche inherited from Volkswagen. The cars raced at Le Mans to a rev limit of 5,000 rpm.

Minus its roof, Porsche 550-02 posed in the Nürburgring paddock before the sports-car race on August 2, 1953. Behind its wheel was Wilhelm Hild, a former DKW racing engineer who was responsible for the 550's design.

In the seven-lap race at the Nürburgring, before the German Grand Prix, Hans Herrmann won driving 550-02. A slot in the nose fed air to an oil cooler, improved in design from one found wanting at Le Mans.

During August 1953 both 550s were overhauled in preparation for their sale to Guatemalan racing enthusiasts. This was 550-02, seen at the Porsche factory before shipment to Central America. It had acquired additional cooling vents along its flanks and a filler cap for a reserve engine-oil supply.

The Guatemalans entered their Porsche coupes in the Mexican Road Race or Carrera Panamericana held in November 1953. While one car retired, the other, driven by Jose Herrarte, won its 1,500-cc category. Both cars remained in Latin America and, so far as is known, no longer exist in their original form.

Porsche was willing to part with its original 550s because it was already building an improved version of which this was the chassis frame, viewed from the top with the front of the car toward the left. In the new layout the rear torsion-bar tube was moved ahead of the engine instead of being behind the transaxle as in the earlier cars. Front suspension was typical VW-type with twin parallel trailing arms and torsion-bar springing.

Two new 550s, of which this was the second, were completed in the late summer of 1953. Posing with 550-04 across the East River from the United Nations was Karl Kling, a Daimler-Benz employee who was one of Germany's leading racers. Porsche designer Erwin Komenda gave the new 550 a strikingly handsome shape.

Karl Kling was unable to complete the second leg of the 1953 Mexican Road Race, in which he is pictured, because his 550's pushrod engine failed. A sister car, driven by Herrmann, suffered a broken steering arm and retired as well, leaving victory to one of the older Guatemalan-entered 550 coupes.

Hans Herrmann returned to Mexico in November 1954 with 550-04. Powered by a new four-cam Porsche engine, he finished third overall and dominated the 1,500-cc class.

Porsche's head of motor sports and public relations, Huschke von Hanstein, took class-winning 550-04 straight from Mexico to Nassau in the Bahamas for its Speed Week in December 1953. This was the car's final outing in the hands of the Porsche company.

Porsche 550-03 was sold after the Mexican race in 1953 and spent its subsequent career in America. One of its owners was Pennsylvanian Evans Hunt, who competed in SCCA events with a pushrod engine. Hunt succumbed to the temptation of the fast downhill bend at Thompson Raceway in Connecticut.

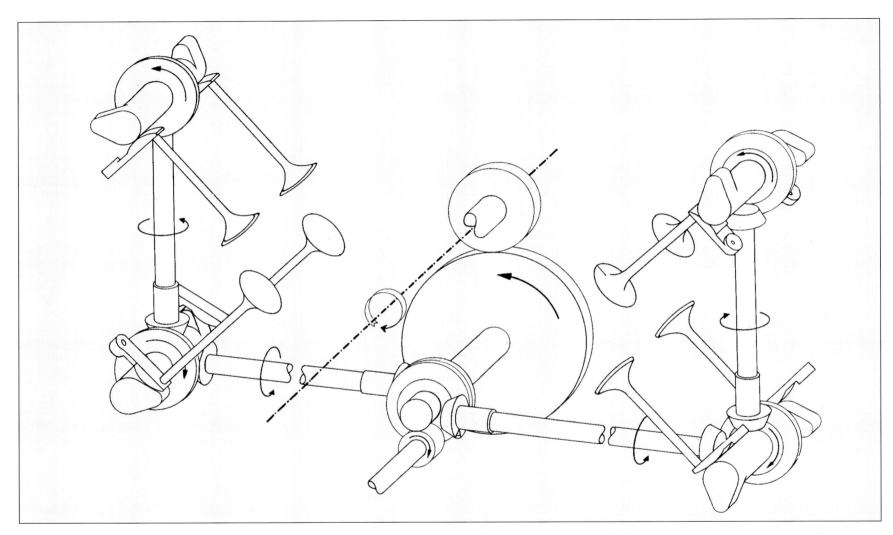

To power its new sports-racing cars Porsche began work in the summer of 1952 on the design of a new flat-four engine. A young engineer, Ernst Fuhrmann, who had shown an aptitude for high-performance design, led the project. He created a system of shafts and bevel gears to drive the twin overhead cams on each cylinder bank, operating inclined overhead valves.

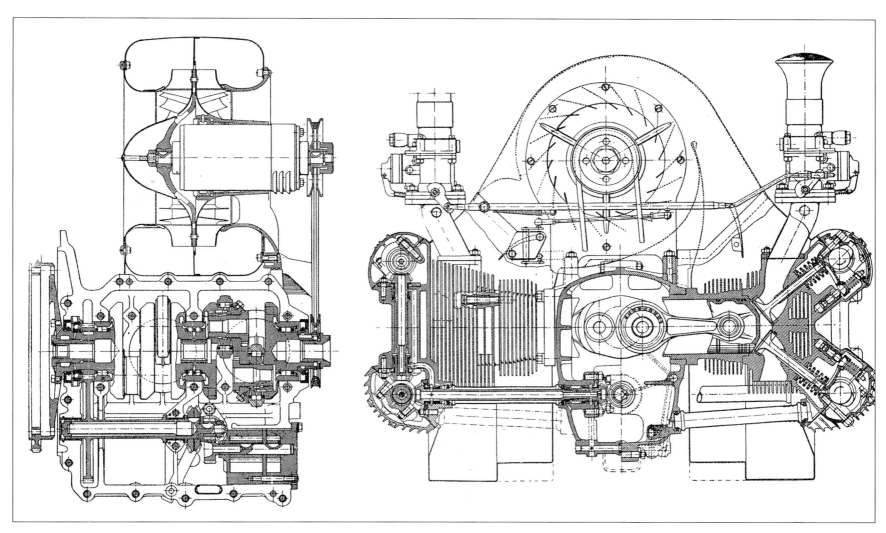

Fuhrmann retained air cooling for his new engine design, the Type 547. He gave ample finning to its cylinder heads and cylinders to cope with the requirements of a high-performance engine. Finger followers were interposed between cam lobes and valve stems with the valves symmetrically inclined at an included angle of 78 degrees.

The Type 547's crankshaft ran on roller bearings for both its three main bearings and its four one-piece connecting-rod big ends. This construction was made possible by the use of a crankshaft built up from separate components by the method developed by Hirth, which used finely threaded bolts and serrated joints to hold the components together. Stuttgart's Hirth made the Type 547 crankshafts for Porsche.

A bottom view of the Type 547 four showed the finning of its dry sump and the very deep cooling fins of its cylinders. The individual cylinders were made by Mahle of aluminum with chrome-plated bores.

The earliest Type 547 engines in 1953 lacked Porsche identification. On each side the inlet camshaft also drove a distributor for the ignition system, which delivered sparks for two plugs per cylinder.

A second-generation version of the Type 547 engine, developed early in 1954, carried the Porsche name on its inlet camshaft covers. Ernst Fuhrmann deliberately designed the engine to be suitable for installation in the road-going Porsche Type 356, and in fact installed an early engine in his own company car.

The first public appearance of Porsche 550-03 and its four-cam engine was in practice for the sports-car race at the Nürburgring on August 2, 1953. A bulge in its rear deck was needed to provide clearance for the new engine's blower housing. Huschke von Hanstein, holding the wheel, gestured to engine designer Ernst Fuhrmann.

The shorter Fuhrmann made a point to von Hanstein next to brand-new 550-03 at the Nürburgring at the beginning of August 1953. Its aluminum body was the work of a new supplier to Porsche, Weinsberg.

With von Hanstein in attendance, Hans Hermann tested the new 550 at the Nürburgring during practice in 1953. Hermann could not match his times with the earlier-model 550, so the new car was not raced.

The first competition appearance for a Porsche 550 powered by the four-cam engine was on August 9, 1953, when veteran Hans Stuck drove it in the famous hillclimb at Freiburg in the Black Forest. Wilhelm Hild, in cap, stood in attendance. Stuck placed third overall.

At the Paris Salon in October 1953 Porsche displayed a new version of its sports-racer that showed it was planning to produce such cars for sale. Built on chassis 550-05, it had rounded nose contours and more refined rear-end lines with purpose-built tail lamps. Quick-detachable racing wheels figured on a Type 550 for the first time.

Shown in company with standard Porsche models at the Paris Salon, the open 550 was a star attraction. Suggestions by Porsche's representatives that the car was ready for early public sale turned out to be premature. This new Porsche model had still to be transformed by several further evolutionary steps.

Its steering wheel was straight from the standard Porsches but otherwise the cockpit of 550-05 shown in Paris in late 1953 was purpose-built. This showed the car after Paris with a different tachometer.

As shown at Paris, Porsche 550-05 was a curious amalgam of stark presentation for racing and more luxurious accoutrements for the road, such as a glove compartment and full windshield complete with wipers. Porsche was still trying to decide what kind of car it should be producing for this market.

Visitors to the Paris Salon in October 1953 were the first to have a look at Porsche's new four-cam Type 547 engine, which had not been revealed during its initial outing at the Nürburgring. Its blower was a remarkably efficient twin-inlet design, housing the generator in its hub. Twin-throat Solex carburetors continued to be used.

The engine in the Paris show car was of the early type without Porsche identification on its camshaft covers. Twin coils for the dual ignition system flanked the transaxle, while the reservoir for the engine's dry-sump oiling system was at the right of the compartment, balancing the battery.

The 1953 Paris Salon prototype, chassis 550-05, remained in Porsche's inventory as a test vehicle. It was given conventional Porsche wheels instead of the knock-off design, which was not proceeded with.

Although decorated with a racing number for promotional purposes, Porsche never entered 550-05 in competition. It nevertheless was an important link in the evolutionary chain that led to the final design of Porsche's 550.

Using a wood model scaled to one-fifth the size of the actual car, Porsche carried out a series of aerodynamic tests on alternative body configurations during 1953. One design with a raised rear deck showed particular promise as a means of reducing drag while retaining an open body shape.

By the end of 1953 a prototype had been built to this promising configuration on chassis number 550-07. With its high rear deck it was nicknamed the *Buckelwagen* or "hunchback car." Like the Porsche shown in Paris it had tapering transparent wind deflectors above its doors.

At the rear of the *Buckelwagen* scale model its design resembled that of the normal 550 but with a higher rear deck that tapered down sharply to its tail. Wind-tunnel tests showed that it offered 16 percent less wind resistance than the same car with a conventional low rear deck and similar full windshield.

The *Buckelwagen* incorporated several chassis features that were ultimately used in the production 550s. One was the passage of the rear frame members below the axle shafts. Another was the use of a one-piece rear deck, which hinged up and back to give full access to the engine and drive train.

The one-piece upward-hinging rear deck of 550-07 attracted press attention when the first *Buckelwagen* was displayed at the Brussels Salon in January 1954. Onlookers grasped that Porsche was still working through its options for the design of this important new model.

A new feature of chassis 550-07 was the positioning of its ignition coils back to back above the clutch housing. This configuration was later adopted for the production 550. The display car was powered by an early Type 547 engine on which the Porsche name was hand-lettered across its inlet-camshaft covers.

Rudy Mailander photographed the *Buckelwagen* on the occasion of its Brussels debut, its angled position highlighting the pleated-leather facing of its raised rear deck. Like the car shown in Paris, this 550 had knock-off wheels.

In March 1954 the *Buckelwagen* was exhibited on the stand of AMAG, the Swiss importer of both VW and Porsche, at the Geneva Salon. Its fanciful surroundings suggested the *joie de vivre* of life in the South of France.

In the lines of its rear deck, Porsche prototype 550-07 was true to the scale model that had inspired it, although its air-inlet grilles were placed more to the rear. Added were rear bumperettes for a modicum of protection.

Although its seats were richly upholstered, the interior of the *Buckelwagen* was otherwise Spartan. To the left of the driver, twin Autopulse electric pumps delivered fuel to the engine from the front-mounted tank.

These rare snapshots show some of the alternatives that Porsche considered to provide a roof for a car of the *Buckelwagen's* configuration. The transparent panel left little in the way of headroom. The helmeted driver was Swiss racer Willy Daetwyler, who was trying the second of two *Buckelwagen* prototypes.

The second *Buckelwagen* was distinguished by a more squarish front edge to its front deck-lid opening. The bulge in its removable roof panel gave a modicum of headroom, at least for the relatively short Hans Herrmann.

Porsche made the first-ever road-racing entry of a Type 550 in the Mille Miglia of May 1954. The car was 550-08, driven by Hans Herrmann accompanied by Herbert Linge, seen here at technical inspection before the start.

In a stunningly successful debut, 550-08 won its 1 1/2-liter class in the Mille Miglia and finished an impressive sixth overall. Here Herrmann and Linge demonstrated it to an appreciative crowd in Germany after the race.

The chassis of the Mille Miglia 550 was an unusual hybrid, having rear frame tubes that passed over its axle shafts. It was fitted with one of the latest engines, with the Porsche name cast into its inlet-cam covers. Italian Weber carburetors replaced the Solexes used hitherto.

The cockpit of the class-winning Type 550 for the Mille Miglia omitted a speedometer—leaving a gaping hole—and added an oil-pressure gauge at the left. Seated astride its steering column was Hans Herrmann's lucky mascot, which certainly did him proud on this occasion.

For the 1954 Le Mans 24-Hour Race, Porsche designed and built four new 550s. The attention to detail in their preparation was exemplified by the way their brake drums were lightened by machining and given transverse drillings through their fins for both cooling and lightness. Rear-suspension trailing arms were drilled for lightness.

Its 1954 Le Mans campaign was by far Porsche's most ambitious effort in racing to date. In jacket and tie Huschke von Hanstein oversaw preparation of the cars, at the rear of one of which gray-haired Wilhelm Hild supervised the efforts of Hubert Mimler, whose skills included test driving as well as car preparation.

The four new 550s built for Le Mans in 1954 carried serial numbers 550-10 through -13. This was 550-11. New features were ventilating louvers in the rear fenders and a pair of driving lights to suit the nocturnal conditions. Color flashes on the rear fenders differed to make the cars easy to identify.

At the rear the new 550s for Le Mans showed the result of steady evolution. They retained rear fenders that were subtly finned to aid directional stability and rear decks that tapered sharply downward. All four exhaust pipes were gathered into a single central outlet that became a 550 hallmark.

Preparation of the Le Mans cars continued in Porsche's garages near the circuit. One mechanic was carrying out an engine change while the other dealt with a rear-brake adjustment. The four-cam engine was famously complex but the rest of the 550 was relatively easy to maintain.

Practice at Le Mans extended late into the evening to give the drivers a chance to adapt their cars and headlamps to night-driving conditions. Chevrolet engineer Zora Arkus-Duntov took part, driving the team's lone 1.1-liter entry. The other three cars had 1 1/2-liter engines.

In 1954 Porsche was consistently using Weber carburetors on its factory racing cars. Screened inlets kept out undesirable foreign objects. A long diagonal tube behind the blower housing provided improved crankcase breathing, solving a problem that had afflicted earlier Type 547 engines.

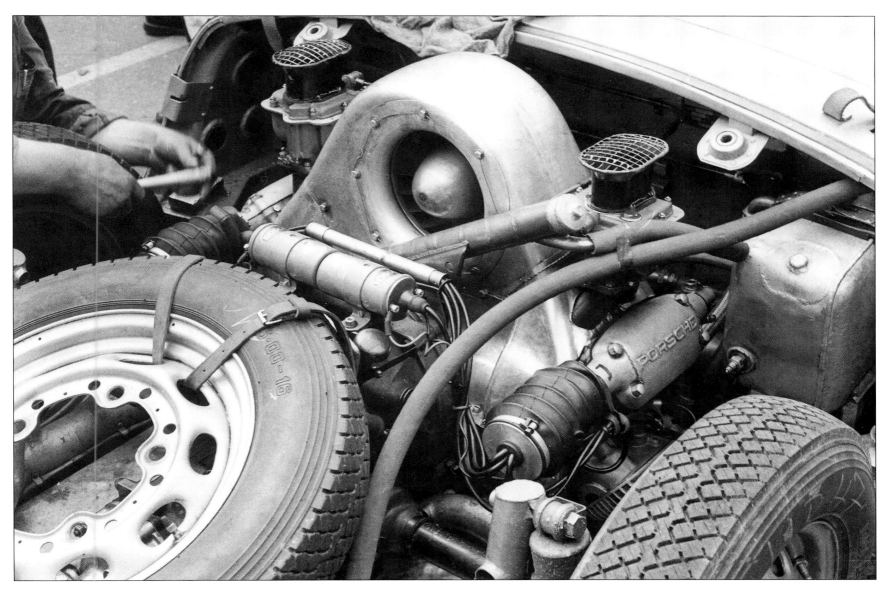

The spare wheel and tire required to be carried at Le Mans were stowed low in the tail above the transaxle. Anticipating rainy conditions, Porsche fitted rubber bellows around the two ignition distributors driven from the inlet camshafts. A hose from the new breather took oil back to the reservoir.

On this Le Mans Porsche one of the protective rubber bellows for the distributor had been pushed back to allow it to be serviced. Lightening efforts even extended to the tube holding the abutment that limited upward rear-suspension travel. A screened scoop supplied cooling air to the rear brakes.

Porsche met the letter of the Le Mans regulations with a rattan passenger seat. Because the rules allowed repairs to be made only with components carried on board, these were stowed in a large wooden box. Wood was also used for the floorboards, as it was less likely to become slippery when spotted with oil.

Racing the length of an entire day at Le Mans can be a lonely business. Zora Arkus-Duntov was partnered by Gustave Olivier in Porsche's 1.1-liter entry, 550-13, competing in its class against Kieft, OSCA and Gordini entries.

With smaller pistons and cylinders to reduce its capacity to 1,089 cc, the Type 547 engine developed 72 bhp at just over 7,000 rpm. Cool weather contributed to a persistent problem with misfiring during the race.

As well as misfiring, the 1.1-liter Porsche entry at Le Mans in 1954 had gear-selection difficulties. These cleared up, however, and the unique Porsche kept running while all its rivals fell by the wayside.

Of Porsche's 1 1/2-liter entries in 1954, 550-12 was the last to be afflicted by the piston problems that caused the two other cars to retire. It was driven by Paul Stasse and Johnny Claes, who nursed their Porsche to the final flag.

During a wet 1954 Le Mans race Porsche's pit staff was kept busy with the servicing requirements of its two remaining entries. Rudy Mailander photographed the 1.1-liter car being serviced late in the race—its drivers Duntov and Olivier conferring—while its bigger-engined sister prepared to return to the fray.

The Claes/Stasse 550-12 finished 12th overall in 1954 and won its class, benefiting from the retirement and disqualification of its rivals. Its last hours were completed on three cylinders after one that had failed was deactivated during two lengthy pit stops. Racing luck was with Porsche this time.

Huschke von Hanstein—festooned as usual with cameras—was delighted at the class win by Duntov and Olivier, the latter sitting behind Duntov at the wheel. Zora's wife Elfi also hitched a ride after the race.

In a study in contrasts the class-winning 1.1-liter Porsche 550 joined the third-place 5.4-liter Cunningham C4R after the finish at Le Mans in 1954. The big American Cunninghams were expressly designed for Le Mans.

Zora Arkus-Duntov was a Chevrolet engineer at the time he raced and won for Porsche at Le Mans in 1954. After the race he filed a detailed technical brief for his employers on the technology of the Porsches and other cars at Le Mans. The inlet to his car's oil cooler was taped over to raise engine temperature to combat its misfiring problem.

Arkus-Duntov was friendly with the Porsche people, so much so that after Le Mans he assisted them in tests of the new 550 that helped significantly to improve its handling. He introduced Porsche to the concept of testing on circular skid pads, making recommendations that led to the productive use of suspension anti-roll torsion bars.

On July 4, 1954 Porsche entered two of its 550s in a 12-hour race at Reims, France. Among the six finishers in their category they placed first and second. The class winner, driven by von Frankenberg and Polensky, was eighth overall.

Huschke von Hanstein piloted one of the Le Mans 550s in the race for 1 1/2-liter sports cars held before the German Grand Prix on August 1, 1954. Porsche's cars swept the first four places, von Hanstein trailing the other three.

Opposition in the race at the Nürburgring on August 1, 1954 included OSCA, Borgward and the East-German EMWs. Nevertheless the young Hans Herrmann, driving number 48, easily led his teammates to victory.

Contributing to the dominance of the Porsches in the Nürburgring sports-car race of 1954 (Herrmann here) were the suspension improvements made after Le Mans, which included a front-mounted torsion anti-roll bar. The engines of 1,498 cc were developing 114 bhp at 6,800 rpm and could be revved safely to 7,500 rpm.

For the curtain-raising sports-car race before the Swiss Grand Prix at Berne on August 22, 1954, the Porsche factory loaned one of its Le Mans cars to Arthur Heuberger, the Swiss BMW importer. Parked on the left in this attractive and eclectic lineup, it won the ten-lap race for 1,600-cc cars.

In 1954 Porsche produced a handful of replicas of its Le Mans 550s for sale to special customers. One such was acquired by Wolfgang Seidel, here racing it in the 1955 Mille Miglia accompanied by Helm Glöckler. They placed eighth overall and easily won their class for 1,500-cc sports cars.

A Le Mans Type 550 sports-racer, complete with its additional driving lights, was acquired by Guatemala-based Jaroslav Juhan. Partnered by Jorge Salas Chaves, he raced it in the 1,000 Kilometers of Buenos Aires on January 23, 1955, a road circuit combining long straights with tight hairpin turns.

Juhan and Chaves wonderfully displayed the capabilities of this new Porsche model in the Buenos Aires race. They dominated their class and finished fourth overall, only two laps behind the winner and on the same lap as a 2-liter Maserati. Porsche had clearly built a car that could be raced not only by the factory but also by its customers.

During 1954 Porsche designed a new version of its 550 to be produced for sale. Its steel tubular frame had added bracing for the transverse rear torsion-bar tube and the towers holding its rear shock absorbers. The view here was from the rear of the car, showing the two jacking points built into the frame's extremities.

Seen from its left rear quarter in the Porsche experimental shop, the frame for the production Type 550 showed its improved rear bracing and a horizontal pad added at the right rear to support the oil reservoir.

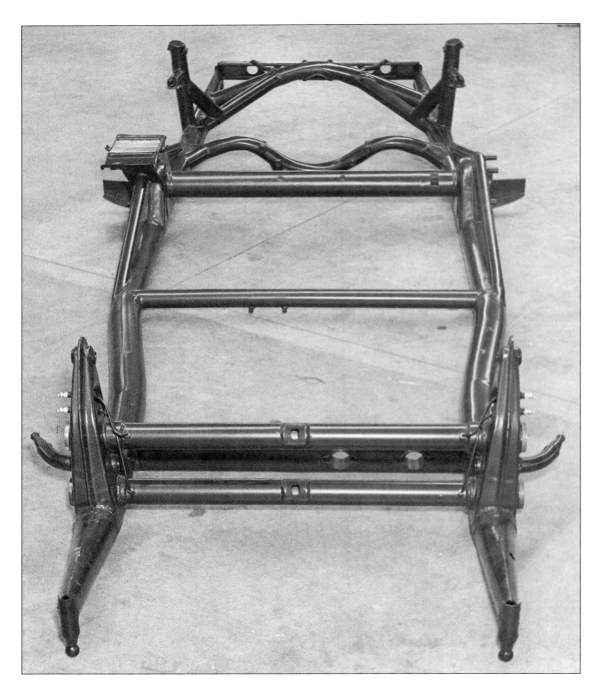

Viewed from the front, the new 550's frame tubes extended forward to the front jacking points. Twin transverse tubes housed the torsion bars (actually bundles of torsion leaves) that provided springing for the trailing-arm front suspension.

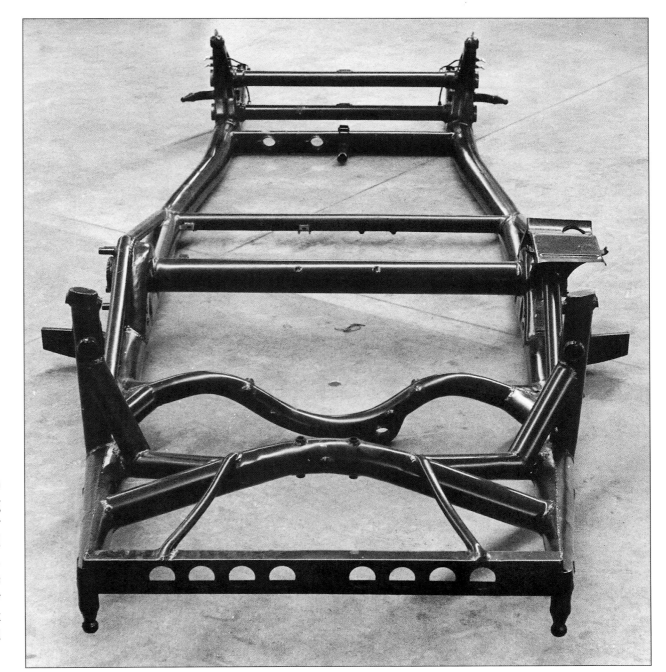

A rear view of the new 550 frame pictured the curving crossmembers at the rear that cradled the engine and suspended the transaxle. Weighing 130 pounds, the 550 frame was eminently serviceable but would not be renowned for its torsional rigidity.

Fichtel & Sachs provided the single-disc clutch for the Type 547 engine with its nine pressure-plate coil springs. Attached by a drilled brace, the ventilator with its angled internal baffles rose up above the clutch housing.

Unique among racing engines, the Type 547 for the series-built Porsche 550 weighed 225 pounds. It was fitted with Solex twin-throat downdraft carburetors, topped by special oval wire-mesh air cleaners made by Knecht.

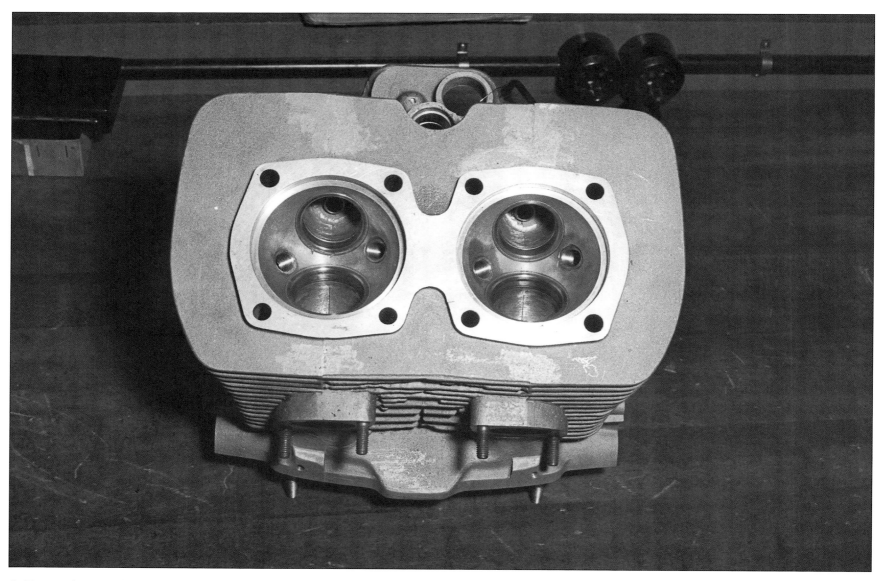

A Type 547 cylinder head showed its hemispherical combustion chambers and apertures for twin spark plugs. Valve-seat inserts in the aluminum head were bronze for the sodium-cooled exhaust valves and stainless steel for the larger and heavier inlet valves. As noted earlier, finning for cooling was generous.

A Type 547 half-crankcase under assembly showed the large low-speed gear that drove the sub-shaft from which spiral bevels turned the shafts driving the cams in each cylinder head. A circular cavity in the sump was ready to accept the pressure and scavenging pumps, a concept carried over from an earlier Porsche design for Cisitalia.

A view under the rear deck of a production 550/1500RS, as the model was named, showed in January 1955 the design features that had evolved from the first such car of 1953. Blower housing was a glossy black.

Distinguishing the series-built Type 547 engines was a bulge at the center of its inlet-camshaft cover to provide additional clearance. This feature was not present on the development engines that had led up to this final design.

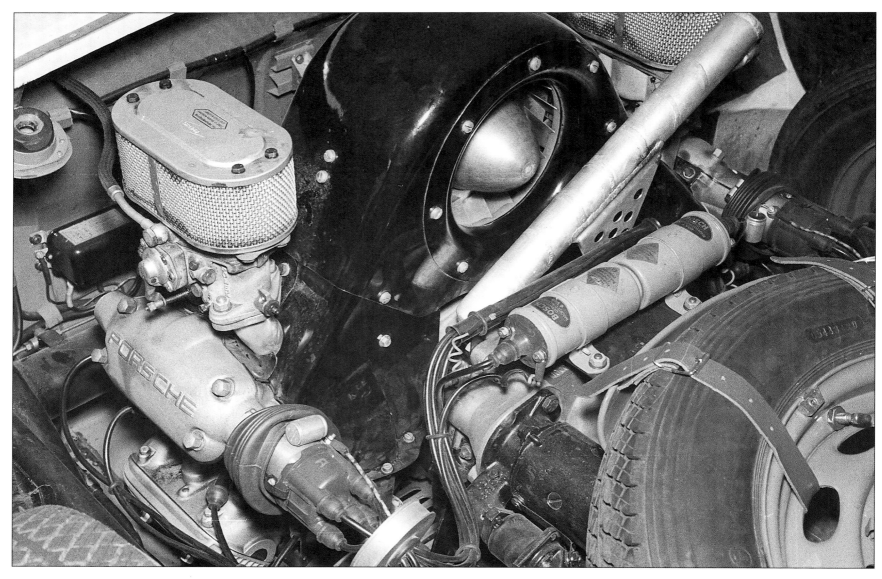

As supplied to Porsche's customers the Type 547's power output was 125 bhp gross and 110 bhp net between 6,500 and 7,000 rpm. Porsche told owners that they could use 7,500 rpm intermittently but recommended a limit of 6,500 rpm for sustained use. Rubber bellows were provided to protect the ignition.

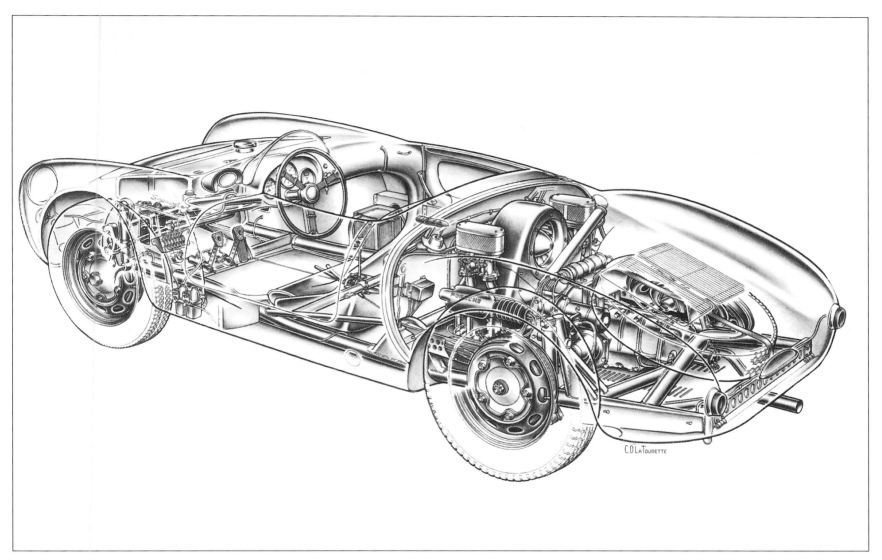

Commissioned by *Sports Cars Illustrated*, Clarence LaTourette's cutaway drawing showed the salient features of the series-built Type 550/1500RS. From its earlier position in the engine bay the battery had migrated to a more forward location opposite the driver. Pedals were floor-hinged and a 24-gallon fuel tank was in the nose.

Rudy Mailander photographed one of the first of the production Porsche 550s on a stretch of Autobahn not far from the Porsche factory in Zuffenhausen, a Stuttgart suburb, in January 1955. Its chromed hubcaps were an incongruous touch on a car that otherwise looked eminently purposeful.

At the suggestion of Max Hoffman, the US importer of Porsches, the new model was given a name as well as a number. Thus "Spyder" script appeared on both sides of its cowl. The term was one that had hitherto been used chiefly in Italy to describe a very light and rudimentary two-seater body style.

Although the series-built Porsche Spyder still had raised rear fenders, these were far less prominent than those of the Le Mans-type cars of 1954. Simple circular tail lamps now sufficed. The cars were more than adequately identified with Porsche names and badges at both front and rear.

The year 1955 was an exciting one for Germany's sports-car enthusiasts with the production launch of both the Mercedes-Benz 300SL and the Porsche 550 Spyder, as it would become familiarly known. That two such great cars should emerge simultaneously in the same city of Stuttgart is one of the great coincidences of motoring history.

A new feature of the production 550 was its rearward-sloping headlamps. Weighing 200 pounds, the aluminum Spyder body was built by Wendler at Reutlingen. Wendler's teardrop emblem appeared behind the front wheels.

The 550 Spyder driver faced a purposeful instrument cluster unchanged in principle from the first cars of almost two years earlier. The tachometer read to 8,000 rpm and the speedometer to 250 km/h, equivalent to 155 mph.

The Autopulse fuel pumps and a battery of a dozen fuses were readily accessible on the left of the 550 Spyder cockpit. A single windscreen and tonneau cover were fitted for events for which a co-driver wasn't needed.

A Spyder in 1956 offered an alternative location for its spare wheel, useful if an extra spare needed to be carried. To the left of the weedy-looking shift lever a simple lift-up latch prevented inadvertent engagement of reverse.

One of the first of the new Spyders to be delivered to its fortunate customer was the property of Switzerland's Hans Stanek. His had a folding top, full windscreen with wiper and hubcaps for its 16-inch aluminum-rimmed wheels.

Hans Stanek would have paid $5,800 for the privilege of becoming a Porsche Spyder owner. In April 1955 he was photographed in action with his new prize at the driving school organized by a Swiss motoring club at Campione.

Also on hand at Campione, an Italian enclave in Switzerland, in April 1955 was a works-presented 550 Spyder. Driven by Porsche tester Paul Denk, the car showed the Porsche colors at a popular event and took part in driver-training exercises as well, just as the second *Buckelwagen* had in 1954.

The works-prepared Spyder driven by Denk at Campione dispensed with both hubcaps and the rubber-faced front bumper strip that was supplied with many production 550s. The car's clean and attractive shape was a tribute to the skills of Porsche's designers, especially Erwin Komenda who had been a Porsche man since the 1930s.

In 1955 Porsche's racing workshop was busy with the preparation of 550 Spyders for customers as well as new racers for its own team. At the right of the row of Spyders was the dark-painted development car of 1953, 550-05.

Proof of what the new Porsche could achieve was provided on March 18, 1955 when Richard von Frankenberg joined Walter Ringgenberg to drive the latter's brand-new Spyder at Montlhéry to attack speed records on the banked French track. They set new international 1,500-cc records at distances and times from 200 miles and 3 hours up to 1,000 kilometers and 6 hours at speeds as high as 131 mph.

In addition to the usual box full of spare parts, preparations for Le Mans in 1955 included a clock attached to the passenger seat. This racing Porsche retained a lever on its steering column for headlamp flashing.

Preparations for Le Mans in 1955 included the checking of spares and tools to be carried aboard the Ecurie Belge entry for Wolfgang Seidel and Olivier Gendebien. They finished an excellent fifth overall and second in class behind another Porsche. Until nightfall, headlamps were well protected.

Driving one of the two 1.1-liter Spyders entered at Le Mans in 1955, Zora Arkus-Duntov co-piloted his to a class victory partnered by Auguste Veuillet, head of Porsche's French importer Sonauto. They were 13th overall and their sister car, entered by his 1954 co-driver Gustave Olivier, was 18th.

Accelerating at the start of the Mulsanne Straight, the Duntov/Veuillet 1.1-liter Porsche lifted its nose in 1955. This was the year of the terrible disaster in which a Mercedes-Benz 300SLR scythed into the crowd. Nevertheless the race was continued and, on this occasion, to Porsche's benefit.

Partnered by Josef Jeser, Gustave Olivier drove his personal 1.1-liter Type 550 Spyder to 18th place overall and second in class at Le Mans in 1955. Their car was the last to complete enough laps to be classified as a finisher.

Of Porsche's team cars for Le Mans in 1955, number 62 was driven to sixth overall and third in class by Helm Glöckler and Jaroslav Juhan. They completed a one-two-three sweep by Porsche of the sports 1,500-cc class.

Winner of the 1 1/2-liter sports-car class at Le Mans in 1955, and victor as well in the important Index of Performance, was factory Spyder number 37 driven by Helmut Polensky and Richard von Frankenberg. In a year of tremendous success at Le Mans for Porsche, they finished fourth overall. New were inlets for brake cooling air.

Special preparation of a 550 Spyder to compete for the Porsche factory in the Grand Prix of Berlin for 1 1/2-liter sports cars on September 25, 1955 included air entries above each carburetor and additional grilles at the extreme rear to reduce overheating that had hampered Spyder performance in the same race a year earlier.

In the Berlin Grand Prix on that city's high-speed AVUS track Richard von Frankenberg faced a strong challenge from the sleek and powerful East German EMWs. After the 155-mile race on September 25, 1955 von Frankenberg was the winner at an average speed of 122.7 mph, in spite of having to round this tight turn at one end of the course.

At the other end of the AVUS track a steep banking allowed progress at undiminished speed. Wind-cheating features of the specially prepared 550 Spyder included blocking off its brake ducts, covering its fuel-filler cap and fitting enclosures for its rear wheels. Its engine was specially tuned to 122 net horsepower.

The 1955 season saw new Porsche Spyders already active in the United States, as in this instance at Cumberland, Maryland. Christiansen's Spyder won its sprint race for 2-liter modified sports cars.

Not all Porsche 550 Spyder adventures ended happily. At Cumberland in 1955 Don McKnought spun off course in his Spyder. Herbert Linge from Stuttgart drove a sister car to third in the race for 1 1/2-liter modified sports cars.

Later Type 550s dispensed with front bumper strips in favor of cooling inlets for their front brakes, like this Spyder racing on an airport circuit at Harewood, Canada in 1956. It was being chased by an Italian OSCA, the Porsche's most persistent rival in its class in both Europe and America.

From the end of 1954 through early 1956 Porsche produced 78 550 Spyders, not including those built for the factory's own use. As late as 1960 well-prepared 550s were still competing in America, as in this instance at Cumberland in 1960. It was being passed by the Type 61 "Birdcage" Maserati of Gaston Andrey.

In 1955 Gordon "Tippy" Lipe installed a pushrod Porsche engine in a modified and re-bodied Formula 3 Cooper chassis to create the purposeful central-seated sports-racer seen to the left of a standard Spyder whose owner had made his own brake-cooling air inlets. In sprint-type American sports-car racing the Lipe "Pooper" and a similar car built by Pete Lovely for West-Coast racing proved highly competitive.

In 1955 at Connecticut's Thompson Raceway a production Porsche 550 Spyder, left, had a touching reunion with one of its earliest ancestors, chassis 550-03, the one in which the four-cam Type 547 engine was first installed. Complete with generous grilles for brake cooling, it was living out its last active years as a racing car. A new generation had proved more than able to carry on.

MORE TITLES FROM ICONOGRAFIX

AMERICAN CULTURE
- COCA-COLA: A HISTORY IN PHOTOGRAPHS 1930-1969 ... ISBN 1-882256-46-8
- COCA-COLA: ITS VEHICLES IN PHOTOGRAPHS 1930-1969 ... ISBN 1-882256-47-6
- PHILLIPS 66 1945-1954 PHOTO ARCHIVE ... ISBN 1-882256-42-5

AUTOMOTIVE
- AMX PHOTO ARCHIVE: FROM CONCEPT TO REALITY ... ISBN 1-58388-062-3
- AUBURN AUTOMOBILES 1900-1936 PHOTO ARCHIVE ... ISBN 1-58388-093-3
- CAMARO 1967-2000 PHOTO ARCHIVE ... ISBN 1-58388-032-1
- CHEVROLET STATION WAGONS 1946-1966 PHOTO ARCHIVE ... ISBN 1-58388-069-0
- CLASSIC AMERICAN LIMOUSINES 1955-2000 PHOTO ARCHIVE ... ISBN 1-58388-041-0
- CORVAIR by CHEVROLET EXP. & PROD. CARS 1957-1969, LUDVIGSEN LIBRARY SERIES ... ISBN 1-58388-058-5
- CORVETTE THE EXOTIC EXPERIMENTAL CARS, LUDVIGSEN LIBRARY SERIES ... ISBN 1-58388-017-8
- CORVETTE PROTOTYPES & SHOW CARS PHOTO ALBUM ... ISBN 1-882256-77-8
- EARLY FORD V-8s 1932-1942 PHOTO ALBUM ... ISBN 1-882256-97-2
- FERRARI-THE FACTORY MARANELLO'S SECRETS 1950-1975, LUDVIGSEN LIB SERIES ... ISBN 1-58388-085-2
- FORD POSTWAR FLATHEADS 1946-1953 PHOTO ARCHIVE ... ISBN 1-58388-080-1
- IMPERIAL 1955-1963 PHOTO ARCHIVE ... ISBN 1-882256-22-0
- IMPERIAL 1964-1968 PHOTO ARCHIVE ... ISBN 1-882256-23-9
- JAVELIN PHOTO ARCHIVE: FROM CONCEPT TO REALITY ... ISBN 1-58388-071-2
- LINCOLN MOTOR CARS 1920-1942 PHOTO ARCHIVE ... ISBN 1-882256-57-3
- LINCOLN MOTOR CARS 1946-1960 PHOTO ARCHIVE ... ISBN 1-882256-58-1
- NASH 1936-1957 PHOTO ARCHIVE ... ISBN 1-58388-086-0
- PACKARD MOTOR CARS 1935-1942 PHOTO ARCHIVE ... ISBN 1-882256-44-1
- PACKARD MOTOR CARS 1946-1958 PHOTO ARCHIVE ... ISBN 1-882256-45-X
- PONTIAC DREAM CARS, SHOW CARS & PROTOTYPES 1928-1998 PHOTO ALBUM ... ISBN 1-882256-93-X
- PONTIAC FIREBIRD TRANS-AM 1969-1999 PHOTO ALBUM ... ISBN 1-882256-95-6
- PONTIAC FIREBIRD 1967-2000 PHOTO HISTORY ... ISBN 1-58388-028-3
- RAMBLER 1950-1969 PHOTO ARCHIVE ... ISBN 1-58388-078-X
- STRETCH LIMOUSINES 1928-2001 PHOTO ARCHIVE ... ISBN 1-58388-070-4
- STUDEBAKER 1933-1942 PHOTO ARCHIVE ... ISBN 1-882256-24-7
- STUDEBAKER HAWK 1956-1964 PHOTO ARCHIVE ... ISBN 1-58388-094-1
- ULTIMATE CORVETTE TRIVIA CHALLENGE ... ISBN 1-58388-035-6

BUSES
- BUSES OF MOTOR COACH INDUSTRIES 1932-2000 PHOTO ARCHIVE ... ISBN 1-58388-039-9
- FAGEOL & TWIN COACH BUSES 1922-1956 PHOTO ARCHIVE ... ISBN 1-58388-075-5
- FLXIBLE TRANSIT BUSES 1953-1995 PHOTO ARCHIVE ... ISBN 1-58388-053-4
- GREYHOUND BUSES 1914-2000 PHOTO ARCHIVE ... ISBN 1-58388-027-5
- MACK® BUSES 1900-1960 PHOTO ARCHIVE* ... ISBN 1-58388-020-8
- PREVOST BUSES 1924-2002 PHOTO ARCHIVE ... ISBN 1-58388-083-6
- TRAILWAYS BUSES 1936-2001 PHOTO ARCHIVE ... ISBN 1-58388-029-1
- TROLLEY BUSES 1913-2001 PHOTO ARCHIVE ... ISBN 1-58388-057-7
- YELLOW COACH BUSES 1923-1943 PHOTO ARCHIVE ... ISBN 1-58388-054-2

EMERGENCY VEHICLES
- AMERICAN AMBULANCE 1900-2002: AN ILLUSTRATED HISTORY ... ISBN 1-58388-081-X
- AMERICAN LAFRANCE 700 SERIES 1945-1952 PHOTO ARCHIVE ... ISBN 1-882256-90-5
- AMERICAN LAFRANCE 700 SERIES 1945-1952 PHOTO ARCHIVE VOLUME 2 ... ISBN 1-58388-025-9
- AMERICAN LAFRANCE 700 & 800 SERIES 1953-1958 PHOTO ARCHIVE ... ISBN 1-882256-91-3
- AMERICAN LAFRANCE 900 SERIES 1958-1964 PHOTO ARCHIVE ... ISBN 1-58388-002-X
- CLASSIC SEAGRAVE 1935-1951 PHOTO ARCHIVE ... ISBN 1-58388-034-8
- CROWN FIRECOACH 1951-1985 PHOTO ARCHIVE ... ISBN 1-58388-047-X
- FIRE CHIEF CARS 1900-1997 PHOTO ALBUM ... ISBN 1-882256-87-5
- HAHN FIRE APPARATUS 1923-1990 PHOTO ARCHIVE ... ISBN 1-58388-077-1
- HEAVY RESCUE TRUCKS 1931-2000 PHOTO GALLERY ... ISBN 1-58388-045-3
- IMPERIAL FIRE APPARATUS 1969-1976 PHOTO ARCHIVE ... ISBN 1-58388-091-7
- INDUSTRIAL AND PRIVATE FIRE APPARATUS 1925-2001 PHOTO ARCHIVE ... ISBN 1-58388-049-6
- LOS ANGELES CITY FIRE APPARATUS 1953-1999 PHOTO ARCHIVE ... ISBN 1-58388-012-7
- MACK MODEL C FIRE TRUCKS 1957-1967 PHOTO ARCHIVE* ... ISBN 1-58388-014-3
- MACK MODEL L FIRE TRUCKS 1940-1954 PHOTO ARCHIVE* ... ISBN 1-882256-86-7
- MAXIM FIRE APPARATUS 1914-1989 PHOTO ARCHIVE ... ISBN 1-58388-050-X
- NAVY & MARINE CORPS FIRE APPARATUS 1836 -2000 PHOTO GALLERY ... ISBN 1-58388-031-3
- PIERRE THIBAULT LTD. FIRE APPARATUS 1918-1990 PHOTO ARCHIVE ... ISBN 1-58388-074-7
- PIRSCH FIRE APPARATUS 1890-1991 PHOTO ARCHIVE ... ISBN 1-58388-082-8
- POLICE CARS: RESTORING, COLLECTING & SHOWING AMERICA'S FINEST SEDANS ... ISBN 1-58388-046-1
- SEAGRAVE 70TH ANNIVERSARY SERIES PHOTO ARCHIVE ... ISBN 1-58388-001-1
- TASC FIRE APPARATUS 1946-1985 PHOTO ARCHIVE ... ISBN 1-58388-065-8
- VOLUNTEER & RURAL FIRE APPARATUS PHOTO GALLERY ... ISBN 1-58388-005-4
- W.S. DARLEY & CO. FIRE APPARATUS 1908-2000 PHOTO ARCHIVE ... ISBN 1-58388-061-5
- WARD LAFRANCE FIRE TRUCKS 1918-1978 PHOTO ARCHIVE ... ISBN 1-58388-013-5
- WILDLAND FIRE APPARATUS 1940-2001 PHOTO GALLERY ... ISBN 1-58388-056-9
- YOUNG FIRE EQUIPMENT 1932-1991 PHOTO ARCHIVE ... ISBN 1-58388-015-1

RACING
- CHAPARRAL CAN-AM RACING CARS FROM TEXAS LUDVIGSEN LIBRARY SERIES ... ISBN 1-58388-066-6
- DRAG RACING FUNNY CARS OF THE 1970s PHOTO ARCHIVE ... ISBN 1-58388-068-2
- EL MIRAGE IMPRESSIONS: DRY LAKES LAND SPEED RACING ... ISBN 1-58388-059-3
- GT40 PHOTO ARCHIVE ... ISBN 1-882256-64-6
- INDY CARS OF THE 1950s, LUDVIGSEN LIBRARY SERIES ... ISBN 1-58388-018-6
- INDY CARS OF THE 1960s, LUDVIGSEN LIBRARY SERIES ... ISBN 1-58388-052-6
- INDIANAPOLIS RACING CARS OF FRANK KURTIS 1941-1963 PHOTO ARCHIVE ... ISBN 1-58388-026-7
- JUAN MANUEL FANGIO WORLD CHAMPION DRIVER SERIES PHOTO ALBUM ... ISBN 1-58388-008-9
- LOST RACE TRACKS TREASURES OF AUTOMOBILE RACING ... ISBN 1-58388-084-4
- MARIO ANDRETTI WORLD CHAMPION DRIVER SERIES PHOTO ALBUM ... ISBN 1-58388-009-7
- MERCEDES-BENZ 300SL RACING CARS 1952-1953 LUDVIGSEN LIBRARY SERIES ... ISBN 1-58388-067-4
- NOVI V-8 INDY CARS 1941-1965 LUDVIGSEN LIBRARY SERIES ... ISBN 1-58388-037-2
- PORSCHE SPYDERS TYPE 550 1953-1956, LUDVIGSEN LIBRARY SERIES ... ISBN 1-58388-092-5
- SEBRING 12-HOUR RACE 1970 PHOTO ARCHIVE ... ISBN 1-882256-20-4
- VANDERBILT CUP RACE 1936 & 1937 PHOTO ARCHIVE ... ISBN 1-882256-66-2

RAILWAYS
- CHICAGO, ST. PAUL, MINNEAPOLIS & OMAHA RAILWAY 1880-1940 PHOTO ARCHIVE ... ISBN 1-882256-67-0
- CHICAGO & NORTH WESTERN RAILWAY 1975-1995 PHOTO ARCHIVE ... ISBN 1-882256-76-X
- GREAT NORTHERN RAILWAY 1945-1970 VOL 2 PHOTO ARCHIVE ... ISBN 1-882256-79-4
- GREAT NORTHERN RAILWAY ORE DOCKS OF LAKE SUPERIOR PHOTO ARCHIVE ... ISBN 1-58388-073-9
- ILLINOIS CENTRAL RAILROAD 1854-1960 PHOTO ARCHIVE ... ISBN 1-58388-063-1
- MILWAUKEE ROAD 1850-1960 PHOTO ARCHIVE ... ISBN 1-882256-61-1
- MILWAUKEE ROAD DEPOTS 1856-1954 PHOTO ARCHIVE ... ISBN 1-58388-040-2
- SHOW TRAINS OF THE 20TH CENTURY ... ISBN 1-58388-030-5
- SOO LINE 1975-1992 PHOTO ARCHIVE ... ISBN 1-882256-68-9
- STREAMLINERS to the TWIN CITIES PHOTO ARCHIVE 400, Twin Zephyrs & Hiawatha Trains ISBN 1-58388-096-8
- TRAINS OF THE TWIN PORTS PHOTO ARCHIVE, DULUTH-SUPERIOR IN THE 1950s .. ISBN 1-58388-003-8
- TRAINS OF THE CIRCUS 1872-1956 ... ISBN 1-58388-024-0
- TRAINS of the UPPER MIDWEST PHOTO ARCHIVE STEAM & DIESEL in the 1950s & 1960s . ISBN 1-58388-036-4
- WISCONSIN CENTRAL LIMITED 1987-1996 PHOTO ARCHIVE ... ISBN 1-882256-75-1
- WISCONSIN CENTRAL RAILWAY 1871-1909 PHOTO ARCHIVE ... ISBN 1-882256-78-6

TRUCKS
- AUTOCAR TRUCKS 1950-1987 PHOTO ARCHIVE ... ISBN 1-58388-072-0
- BEVERAGE TRUCKS 1910-1975 PHOTO ARCHIVE ... ISBN 1-882256-60-3
- BROCKWAY TRUCKS 1948-1961 PHOTO ARCHIVE* ... ISBN 1-882256-55-7
- CHEVROLET EL CAMINO PHOTO HISTORY INCL GMC SPRINT & CABALLERO ... ISBN 1-58388-044-5
- CIRCUS AND CARNIVAL TRUCKS 1923-2000 PHOTO ARCHIVE ... ISBN 1-58388-048-8
- DODGE B-SERIES TRUCKS RESTORER'S & COLLECTOR'S REFERENCE GUIDE & HISTORY ISBN 1-58388-087-9
- DODGE PICKUPS 1939-1978 PHOTO ALBUM ... ISBN 1-882256-82-4
- DODGE POWER WAGONS 1940-1980 PHOTO ARCHIVE ... ISBN 1-882256-89-1
- DODGE POWER WAGON PHOTO HISTORY ... ISBN 1-58388-019-4
- DODGE RAM TRUCKS 1994-2001 PHOTO HISTORY ... ISBN 1-58388-051-8
- DODGE TRUCKS 1929-1947 PHOTO ARCHIVE ... ISBN 1-882256-36-0
- DODGE TRUCKS 1948-1960 PHOTO ARCHIVE ... ISBN 1-882256-37-9
- FORD 4X4s 1935-1990 PHOTO HISTORY ... ISBN 1-58388-079-8
- FORD HEAVY-DUTY TRUCKS 1948-1998 PHOTO HISTORY ... ISBN 1-58388-043-7
- FREIGHTLINER TRUCKS 1937-1981 PHOTO ARCHIVE ... ISBN 1-58388-090-9
- JEEP 1941-2000 PHOTO ARCHIVE ... ISBN 1-58388-021-6
- JEEP PROTOTYPES & CONCEPT VEHICLES PHOTO ARCHIVE ... ISBN 1-58388-033-X
- MACK MODEL AB PHOTO ARCHIVE* ... ISBN 1-882256-18-2
- MACK AP SUPER-DUTY TRUCKS 1926-1938 PHOTO ARCHIVE* ... ISBN 1-882256-54-9
- MACK MODEL B 1953-1966 VOL 2 PHOTO ARCHIVE* ... ISBN 1-882256-34-4
- MACK EB-EC-ED-EE-EF-EG-DE 1936-1951 PHOTO ARCHIVE* ... ISBN 1-882256-29-8
- MACK EH-EJ-EM-EQ-ER-ES 1936-1950 PHOTO ARCHIVE* ... ISBN 1-882256-39-5
- MACK FC-FCSW-NW 1936-1947 PHOTO ARCHIVE* ... ISBN 1-882256-28-X
- MACK FG-FH-FJ-FK-FN-FP-FT-FW 1937-1950 PHOTO ARCHIVE* ... ISBN 1-882256-35-2
- MACK LF-LH-LJ-LM-LT 1940-1956 PHOTO ARCHIVE* ... ISBN 1-882256-38-7
- MACK TRUCKS PHOTO GALLERY* ... ISBN 1-882256-88-3
- NEW CAR CARRIERS 1910-1998 PHOTO ALBUM ... ISBN 1-882256-98-0
- PLYMOUTH COMMERCIAL VEHICLES PHOTO ARCHIVE ... ISBN 1-58388-004-6
- REFUSE TRUCKS PHOTO ARCHIVE ... ISBN 1-58388-042-9
- RVs & CAMPERS 1900-2000: AN ILLUSTRATED HISTORY ... ISBN 1-58388-064-X
- STUDEBAKER TRUCKS 1927-1940 PHOTO ARCHIVE ... ISBN 1-882256-40-9
- WHITE TRUCKS 1900-1937 PHOTO ARCHIVE ... ISBN 1-882256-80-8

TRACTORS & CONSTRUCTION EQUIPMENT
- CASE TRACTORS 1912-1959 PHOTO ARCHIVE ... ISBN 1-882256-32-8
- CATERPILLAR PHOTO GALLERY ... ISBN 1-882256-70-0
- CATERPILLAR POCKET GUIDE THE TRACK-TYPE TRACTORS 1925-1957 ... ISBN 1-58388-022-4
- CATERPILLAR D-2 & R-2 PHOTO ARCHIVE ... ISBN 1-882256-99-9
- CATERPILLAR D-8 1933-1974 PHOTO ARCHIVE INCLUDING DIESEL 75 & RD-8 ... ISBN 1-882256-96-4
- CATERPILLAR MILITARY TRACTORS VOLUME 1 PHOTO ARCHIVE ... ISBN 1-882256-16-6
- CATERPILLAR MILITARY TRACTORS VOLUME 2 PHOTO ARCHIVE ... ISBN 1-882256-17-4
- CATERPILLAR SIXTY PHOTO ARCHIVE ... ISBN 1-882256-05-0
- CATERPILLAR TEN PHOTO ARCHIVE INCLUDING 7C FIFTEEN & HIGH FIFTEEN ... ISBN 1-58388-011-9
- CATERPILLAR THIRTY PHOTO ARCHIVE 2ND ED. INC. BEST THIRTY, 6G THIRTY & R-4 ISBN 1-58388-006-2
- CIRCUS & CARNIVAL TRACTORS 1930-2001 PHOTO ARCHIVE ... ISBN 1-58388-076-3
- CLETRAC AND OLIVER CRAWLERS PHOTO ARCHIVE ... ISBN 1-882256-43-3
- CLASSIC AMERICAN STEAMROLLERS 1871-1935 PHOTO ARCHIVE ... ISBN 1-58388-038-0
- FARMALL CUB PHOTO ARCHIVE ... ISBN 1-882256-71-9
- FARMALL F-SERIES PHOTO ARCHIVE ... ISBN 1-882256-02-6
- FARMALL MODEL H PHOTO ARCHIVE ... ISBN 1-882256-03-4
- FARMALL MODEL M PHOTO ARCHIVE ... ISBN 1-882256-15-8
- FARMALL REGULAR PHOTO ARCHIVE ... ISBN 1-882256-14-X
- FARMALL SUPER SERIES PHOTO ARCHIVE ... ISBN 1-882256-49-2
- FORDSON 1917-1928 PHOTO ARCHIVE ... ISBN 1-882256-33-6
- HART-PARR PHOTO ARCHIVE ... ISBN 1-882256-08-5
- HOLT TRACTORS PHOTO ARCHIVE ... ISBN 1-882256-10-7
- INTERNATIONAL TRACTRACTOR PHOTO ARCHIVE ... ISBN 1-882256-48-4
- JOHN DEERE MODEL A PHOTO ARCHIVE ... ISBN 1-882256-12-3
- JOHN DEERE MODEL D PHOTO ARCHIVE ... ISBN 1-882256-00-X
- MARION CONSTRUCTION MACHINERY 1884-1975 PHOTO ARCHIVE ... ISBN 1-58388-060-7
- MARION MINING & DREDGING MACHINES PHOTO ARCHIVE ... ISBN 1-58388-088-7
- OLIVER TRACTORS PHOTO ARCHIVE ... ISBN 1-882256-09-3
- RUSSELL GRADERS PHOTO ARCHIVE ... ISBN 1-882256-11-5
- TWIN CITY TRACTOR PHOTO ARCHIVE ... ISBN 1-882256-06-9

*THIS PRODUCT IS SOLD UNDER LICENSE FROM MACK TRUCKS, INC. MACK IS A REGISTERED TRADEMARK OF MACK TRUCKS, INC. ALL RIGHTS RESERVED.

All Iconografix books are available from direct mail specialty book dealers and bookstores worldwide, or can be ordered from the publisher. For book trade and distribution information or to add your name to our mailing list and receive a **FREE CATALOG** contact:

Iconografix, PO Box 446, Dept BK, Hudson, Wisconsin, 54016 Telephone: (715) 381-9755, (800) 289-3504 (USA), Fax: (715) 381-9756

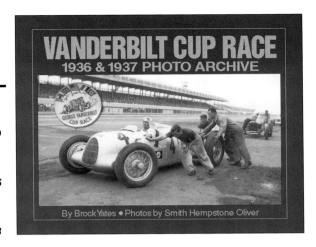

MORE GREAT BOOKS FROM ICONOGRAFIX

GT40 PHOTO ARCHIVE
ISBN 1-882256-64-6

VANDERBILT CUP RACE 1936 & 1937 PHOTO ARCHIVE
ISBN 1-882256-66-2

CHAPARRAL CAN-AM RACING CARS FROM TEXAS LUDVIGSEN LIBRARY SERIES
ISBN 1-58388-066-6

MERCEDES-BENZ 300SL RACING CARS 1952-1953 LUDVIGSEN LIBRARY SERIES
ISBN 1-58388-067-4

INDIANAPOLIS RACING CARS OF FRANK KURTIS 1941-1963 PHOTO ARCHIVE
ISBN 1-58388-026-7

LOST RACE TRACKS TREASURES OF AUTOMOBILE RACING
ISBN 1-58388-084-4

FERRARI-THE FACTORY MARANELLO'S SECRETS 1950-1975 LUDVIGSEN LIBRARY SERIES
ISBN 1-58388-085-2

ICONOGRAFIX, INC.
P.O. BOX 446, DEPT BK,
HUDSON, WI 54016
FOR A FREE CATALOG CALL:
1-800-289-3504

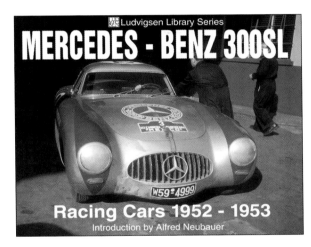
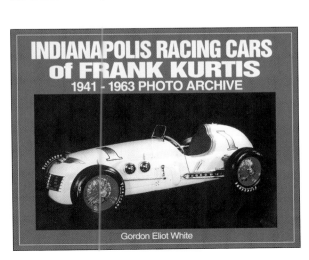
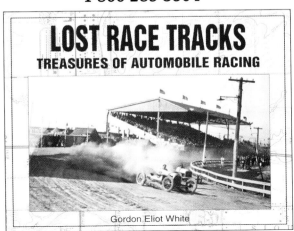

LUDVIGSEN LIBRARY

Those photos in this book that are the copyright holdings of the Ludvigsen Library are available for purchase by enthusiasts. They make wonderful gifts and decorations. The Library, founded and owned by Karl Ludvigsen, is one of the world's most extensive repositories of reference material about cars and the motor industry. Specializing in car and motor-racing photography, the Library holds much rare and unpublished original material by John Dugdale, Edward Eves, Max Le Grand, Peter Keen, Karl Ludvigsen, Rodolfo Mailander, Ove Nielsen, Stanley Rosenthall and others.

All black and white prints are hand finished to museum display standards using the finest Ilford 1K fibre which gives a beautiful, durable finish that is perfect for mounting and display. Prints can be ordered from the Ludvigsen Library at the address below in three sizes at the following prices:

10 x 12 inches	US$40.00	UK£25.00
12 x 16 inches	US$55.00	UK£35.00
16 x 20 inches	US$75.00	UK£45.00

Please inquire concerning color, other sizes, and other subjects. Prices do not include packing and shipping fees, which will be advised in advance.

LUDVIGSEN LIBRARY: SCOLES GATE, REDE ROAD, HAWKEDON, SUFFOLK, IP29 4AU, UNITED KINGDOM
TELEPHONE & FACSIMILE +44 (01284) 789246
E-MAIL KARLCARS@BTINTERNET.COM   HTTP://WWW.LUDVIGSEN.COM